RICE

RICE

by Sylvia A. Johnson

Photographs by Noboru Moriya

A Lerner Natural Science Book

Lerner Publications Company ▪ Minneapolis

Sylvia A. Johnson, Series Editor

Translation of original text by Wesley M. Jacobsen

The publisher wishes to thank Kristen O'Brien, Rice Council for Market Development, and Roberta Sladky, Department of Botany, University of Minnesota, for their assistance in the preparation of this book.

Additional photographs provided by: p. 6, Judy Frater; p. 9, 20, 40 (right), 43, 44, Rice Council for Market Development; p. 23 (right), Susumu Kanozawa; p. 37, Hiroo Koike. Drawing on p. 15 by Yooji Watanabe.

The glossary on page 45 gives definitions and pronunciations of words shown in **bold type** in the text.

LIBRARY OF CONGRESS CATALOGING IN PUBLICATION DATA

Johnson, Sylvia A.
  Rice.

  (A Lerner Natural Science book)
  Adaptation of: Ine no isshō/Noboru Moriya.
  Includes index.
  Summary: An introduction to rice, a member of the grass family, discussing how it is grown and cultivated as a primary food crop for one half of the world's peoples. Includes a glossary of terms.
  1. Rice—Juvenile literature. [1. Rice] I. Moriya, Noboru, ill. II. Moriya, Noboru. Ine no isshō. III. Title. IV. Series.
  SB191.R5J53   1985   633.1'8   85-19754
  ISBN 0-8225-1466-4 (lib. bdg.)

International Standard Book Number: 0-8225-1466-4
Library of Congress Catalog Number: 85-19754

1 2 3 4 5 6 7 8 9 10 94 93 92 91 90 89 88 87 86 85

About half the people in the world depend on one food—rice—for their basic nourishment. In China, Japan, India, and many other parts of Asia, rice is eaten at almost every meal. This grain is so important that in some Asian languages, including Japanese, the word for "rice" is the same as the word meaning "food." The countries of Asia devote more than 300 million acres of land to the cultivation of the lush green plants that produce the little white grains.

An Indian woman working in a rice field. India and China are the leading rice-producing countries in the world.

Very special conditions are required to produce the 350 million tons of rice that are grown each year throughout the world. Rice plants need warm temperatures and plenty of moisture to develop. The best average temperature for a good rice crop is at least 75 degrees Fahrenheit (24 degrees Centigrade). Rice plants need so much moisture that in most areas, they are grown in fields flooded with water.

These requirements are quite different from those of wheat, another grain that, in the form of bread, feeds many millions of people. Wheat grows in generally dry regions with temperate

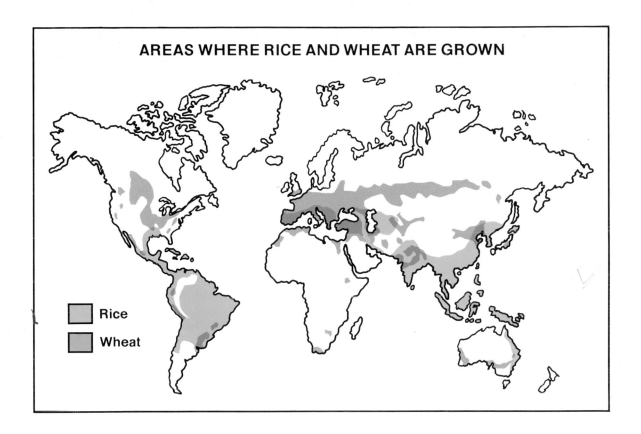

AREAS WHERE RICE AND WHEAT ARE GROWN

Rice
Wheat

climates. Both wheat and rice are grown in North America, but wheat is a much more important crop covering many acres of American soil.

Wheat, rice, and other cereal grains like oats and barley all belong to the family of grass plants, known by the scientific name Gramineae. Another important food crop in this group is corn. All these grain-producing plants are closely related to the grass that grows on lawns and prairies as well as to the bamboo plants of tropical regions.

Let's take a close look at the member of the grass family that produces the tiny white kernels we know as rice.

Good seeds are important in producing healthy rice plants. The seed shown in the photograph at the bottom seems to be a good one because it has a full supply of endosperm, a material that will nourish the growth of the new plant. The seed at the top is lacking in this vital substance.

There is only one species of cultivated rice—*Oryza sativa*—but there are almost 8,000 varieties of rice within this group. Some kinds of plants produce long-grain rice, whereas others produce medium or short grains. There are also varieties of scented or aromatic rices that have a special flavor of their own. Many kinds of rice have been specially bred to suit growing conditions in particular regions. Growers must choose among all these types when deciding on the kind of rice to plant in their fields.

As with most food crops, the cultivation of rice begins with planting seeds. The "seeds" sown to produce new rice plants are, in fact, grains of rice, very similar to the rice grains that are eaten for food. (We will find out later that the grain and the seed of a rice plant are not quite the same in botanical terms.)

After rice seeds are soaked in water to soften them, they are ready to be planted. In many Asian countries, rice seeds are usually sown by hand in special nursery beds (*below*). In the United States, growers often use low-flying airplanes to broadcast the seeds over flooded fields (*right*).

A rice seed begins to sprout.

Buried in the warm, muddy soil, the rice seeds are soon ready to begin their development. Contained within each seed are all the parts needed for the growth of a new rice plant. Packed into one end of the seed is the **embryo**, the tiny plant-to-be. The embryo contains the developing root, stem, and leaves of the new plant. It also includes a special structure called a **cotyledon**, which will play an important role in the growth of the plant.

The rest of the seed is almost completely filled with **endosperm**. This milky white substance is loaded with starch and other nutrients that will feed the developing plant until it is ready to produce its own food. The nutrients will be digested by the cotyledon and transferred to the growing embryo.

All these parts of the rice seed are enclosed within several different kinds of coats or coverings. On the outside of the seed is a tough **hull**, and under it is a **seed coat** as well as several other layers that protect the seed's precious contents. When the seed sprouts, the developing root, stem, and leaves emerge through an opening in these coverings.

10

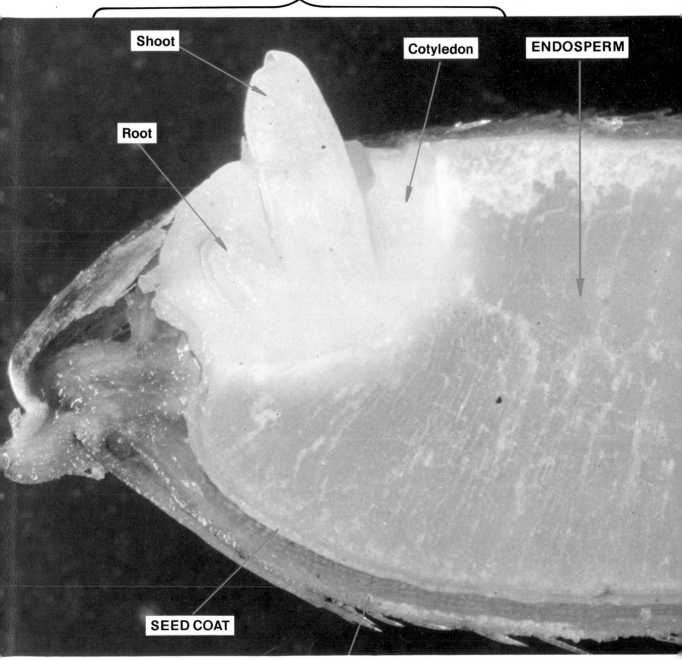

EMBRYO

Shoot

Cotyledon

ENDOSPERM

Root

SEED COAT

HULL

11

LEAF

COLEOPTILE

ROOT

In the first stage of the rice plant's growth, the tiny root pushes its way down into the soil while the **shoot** (the developing leaves and stem) breaks through the surface and reaches up toward the sun. The shoot is covered by a cone-like structure called a **coleoptile**, which protects its tender tip from injury. As the young plant continues to grow, the coleoptile withers and the first long, pointed leaf appears. During this early period of development, the seed, with the cotyledon and endosperm inside it, remains buried in the soil.

The early development of a rice plant reveals some interesting facts about the relationship of rice to other members of the plant kingdom. Rice and all other grasses belong to a group of plants known as **monocots** because they have only one cotyledon.

The first leaves of corn, a monocot plant, are true leaves (*left*). The morning glory, a dicot, produces two seed leaves, or cotyledons (*right*), before its true leaves develop.

Many other kinds of plants, including roses, morning glories, and such familiar vegetables as beans and peas, are **dicots**. These plants have two cotyledons and usually lack endosperm; the food supply for the embryo is stored in the cotyledons themselves.

When a dicot such as a morning glory sprouts, its two cotyledons emerge from the soil. They look like green leaves and, in fact, are known as seed leaves. In monocot plants like rice and corn, the first leaf to make its way into the sunlight is a true leaf, not a seed leaf. The single cotyledon does not emerge but remains underground inside the seed coat, digesting the nutrients in the endosperm and transferring them to the growing young plant.

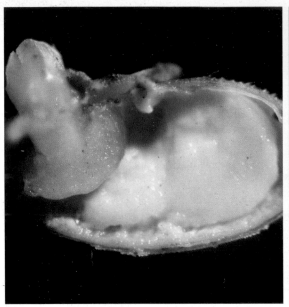

Left: When a rice seed sprouts, it is filled with thick white endosperm. *Right:* Fifteen days later, the seed is shrunken and hollow, its endosperm used as food by the developing plant.

Whether a plant is a monocot like rice and corn or a dicot like the morning glory and pea, its early development follows the same basic pattern. The young plant is nourished by means of the cotyledon until its leaves are able to take over the job of providing food.

Rice and other grasses have green leaves like all plants, but they look somewhat different from the leaves of morning glories, roses, peas, and other dicot plants. Grass leaves are usually long and thin, and they grow in a special way. Each leaf is made up of two main parts, a **sheath** and a **blade**. The sheath is the lower part of the leaf, and it grows wrapped around the plant stem, as you can see in the drawing on the opposite page. The leaf blade, connected to the top of the sheath, is flat and pointed.

14

Below: The parts of a rice leaf. *Right:* Beaded with moisture, the slender leaf blades of rice plants gleam in the sun.

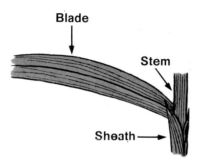

Blade

Stem

Sheath

As the young rice plant grows, it develops new leaves from points on the plant stem called **nodes**. Each new leaf grows higher up on the stem and on the opposite side from the leaf that appeared before it.

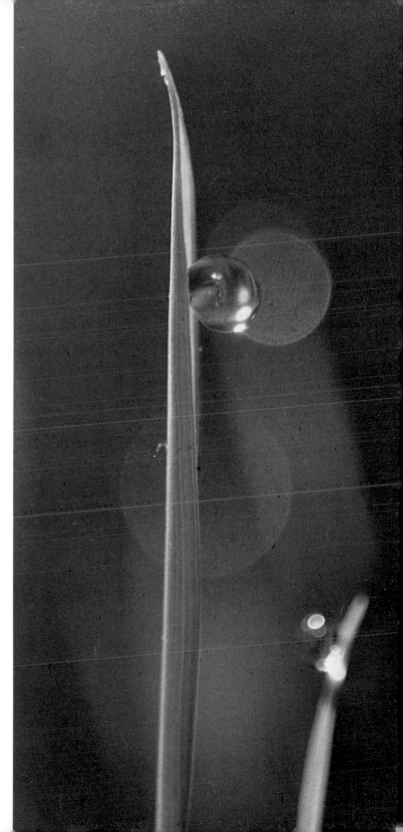

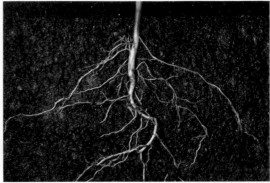

Left: The fibrous root system of a young rice plant begins to develop. *Above:* The tap root of a morning glory is typical of dicot plants.

At the same time that the rice plant produces leaves above the ground, it is also developing beneath the surface of the soil. Its roots expand and dig into the earth, providing a system of support and taking in moisture and nutrients needed for growth.

Although they play the same role in the life of the plant, the roots of rice plants and other monocots have a different structure than the roots of dicot plants. A morning glory has one main root, called a **tap root**, that develops from the plant embryo. Many smaller secondary roots eventually grow from this thick root.

Rice, like all members of the grass family, produces a system of **fibrous roots**. The first tiny root that grows from the embryo is soon joined by many other fine roots that spring from the lower part of the stem. This mass of roots gives rice and other grasses great stability and enables them to grow in soil where dicot plants could not survive.

Like all grass plants, rice has a mass of fine roots instead of one main root.

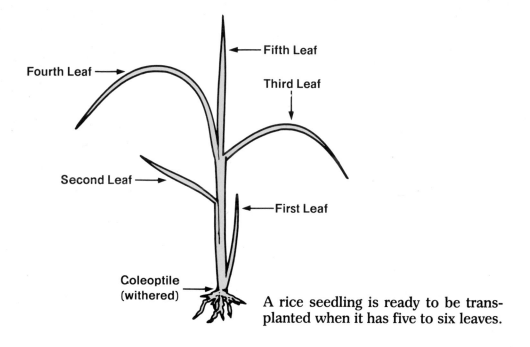

Fifth Leaf

Fourth Leaf

Third Leaf

Second Leaf

First Leaf

Coleoptile
(withered)

A rice seedling is ready to be transplanted when it has five to six leaves.

The early stage of a rice plant's development lasts for about 30 to 40 days. At the end of this time, the plant has produced five or six long, slender leaves. It is now a **seedling**, a young plant that is ready to begin its period of vigorous growth.

In most rice-growing countries in Asia, the seedlings have been developing in nursery beds, their bright green leaves forming dense masses of plants surrounded by water. The time has now come, however, to move them to the **paddies**, the flooded fields where they will continue their development.

In the United States and some other western countries where rice is grown, nursery beds are not usually used. The rice is planted directly in the paddy fields, where it grows from seed to mature plant.

18

Rice seedlings in a nursery bed

The dark lines of levees criss-cross a field of young rice plants.

Paddies have to be specially prepared for the growing of rice. Most paddies are surrounded by **dikes** or **levees**, low walls of dirt that keep the water from running off the flooded fields. To prepare large fields on level land, rice farmers use earth-moving equipment to build up dikes and create individual paddies enclosed by dirt walls. In areas where rice is grown in mountainous regions, small terraced paddies are cut into the hillsides like steps.

Workers transplanting rice seedlings in a flooded paddy. Two or three seedlings are planted in a clump, forming neat rows of healthy young plants.

The leaves of dicot plants (*left*) have a net-like pattern of veins, while monocot leaves (*right*) have parallel veins.

After the rice seedlings are moved to the paddies, they grow rapidly, producing more and more green leaves. These slender leaves are essential to the life of a rice plant because they provide the nourishment that allows the plant to grow.

Like the leaves of all green plants, the rice plant's leaves make food through the process of **photosynthesis**. They contain the green pigment chlorophyll, which, when exposed to sunlight, combines water and the gas carbon dioxide to make glucose, a kind of sugar.

The food-processing centers in a plant's leaves depend on a complicated transportation system made up of tiny veins. These veins are found both in the leaves and in the plant stem. Some carry water from the roots to the food-making cells in the leaves.

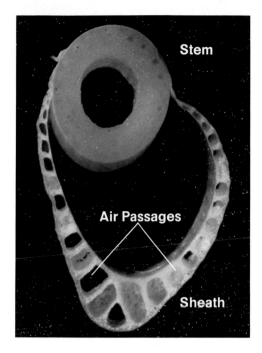

Stem

Air Passages

Sheath

Left: A cross-section of the stem and the attached leaf sheath of a rice plant. As in most monocot plants, the stem is hollow. The veins that run through it are clustered in bundles that are scattered throughout the stem tissue. In the sheath can be seen the large air cavities that are typical of rice plant leaves.

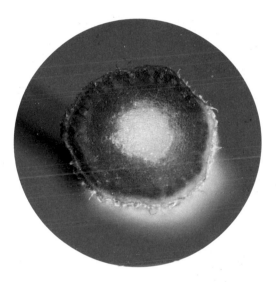

Right: The stem of a sunflower has the characteristics common to dicot stems. It is solid rather than hollow, and its bundles of veins are arranged in a circular pattern around the outside edge of the stem. (In this photograph, a red dye was used to show the vein bundles.)

Others transport glucose to all parts of the plant, where it is used as a fuel for growth.

The system of veins found in rice and other grasses provides another example of the differences between these monocot plants and their dicot relatives. The leaves of dicots have a net-like pattern of veins, but the veins of monocot leaves are arranged in long, parallel rows. There are also some basic differences in the stems of monocots and dicots, as you can see in the photographs above.

Absorbing energy from the hot sun and moisture from the water in which they stand, young rice plants grow vigorously.

Left: New stems called tillers develop from a rice plant's original stem. *Right:* This cross-section photograph shows the relationship among a plant's tillers.

Rice plants grow not only by producing new leaves but also by adding whole new stems. This method of growth is typical of many kinds of grass plants. It is known as **tillering**, and the new stems are called **tillers**.

Tillers develop not from the plant roots but from the base of the original stem. New shoots spring up from nodes in this area and eventually become separate stems with their own leaves. As the rice plant grows, other tillers develop from the first generation of new stems so that the plant is eventually made up of a cluster of stems all connected to the same root system. (The cross-section photograph above shows how the stems of a rice plant are joined to each other.)

In some Asian countries, rice paddies are weeded by hand to keep them free of undesirable plants.

The development of rice, like that of all crops grown for food, can be threatened by pests and diseases that attack the plants. Weeds can be a serious problem in rice paddies. They take up space needed by the rice plants and deprive them of sunlight and air. Aquatic plants such as duckweed cover the surface of the water and keep its temperature lower than suitable for the growth of rice.

Other enemies of rice are insects that attack the plants in various stages of their development. Stem borers are moth larvae that eat their way into the inside of the plant stems, causing great damage. Army worms eat stems and leaves, while rice stinkbugs prefer the tender grains of rice before they harden.

To combat insects and weeds and to prevent devastating plant diseases like blast, growers usually spray their fields with various kinds of chemicals. They also try to use varieties of rice that will produce plants resistant to such problems.

26

Arum (*right*) and duckweed (*below left*) are two of the weeds that can cause problems in rice paddies. The stem borer (*below right*) damages rice plants by burrowing into their stems.

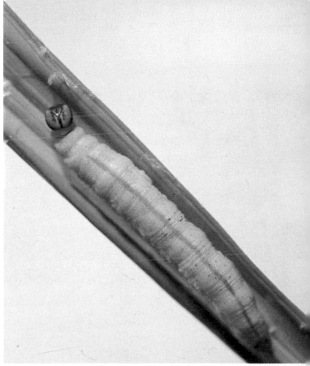

A snowy heron catches a frog in a paddy of mature rice plants.

If the rice plants are not injured by pests or disease, they will reach their full growth about three months from the time of planting. Mature rice plants have many strong stems and healthy green leaves. They can grow to a height of two to six feet (60 to 180 centimeters), depending on the variety.

After reaching their full size, the plants are ready to enter a new period of growth—the reproductive stage of their development. At the beginning of this stage, a new part of the rice plant starts to grow. Deep within each stem, structures known as **heads** have begun to develop. These heads or **panicles** are very important because they contain the rice plant's reproductive organs.

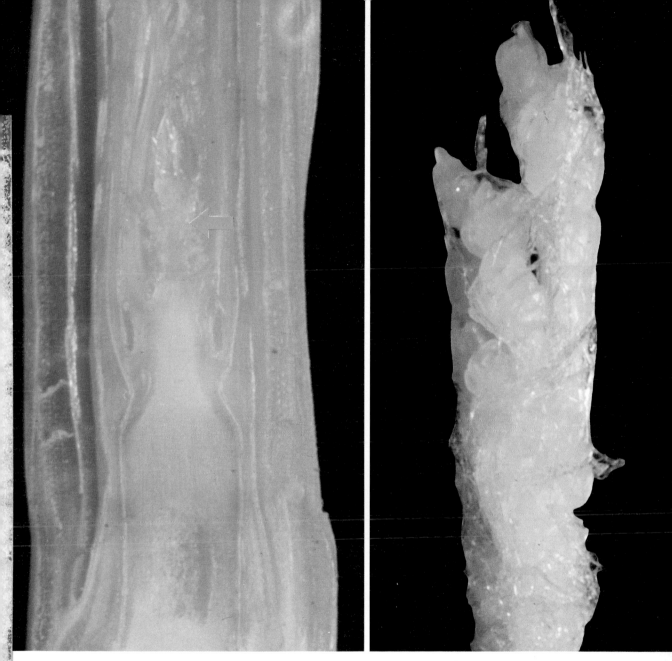

Left: A head (indicated by the arrow) begins to develop within a rice plant stem. *Right:* This photograph shows the head four days later, with the surrounding stem removed. It has grown to about ⅛ inch (3 millimeters) in length.

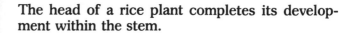

The head of a rice plant completes its development within the stem.

The period in which the head develops is the most crucial in the life of a rice plant. During this time, no new stems or leaves are produced. All nutrients are needed to nourish the developing reproductive part of the plant. If temperatures are too low or there is not enough moisture in this period, the plant's productivity will be seriously affected.

After developing within the stem for about 25 days, the fully developed head emerges at the top, coming out through the sheath of the uppermost leaf. The head or panicle of a rice plant is made up of several primary branches that are further divided into shorter secondary branches. On the secondary branches are small stalks called **spikelets,** each of which bears a single rice flower. (The drawing on the opposite page shows the structure of a rice panicle, which is difficult to see in a photograph.)

Flower

Spikelet

PANICLE

These heads have emerged at the tops of the plant stems.

THE PARTS OF A COMPLETE FLOWER

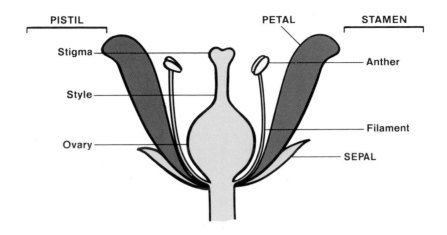

The flowers of rice and other grasses don't look very much like the large, brightly colored flowers of roses, morning glories, or sunflowers. They are quite small, and they lack some of the parts of these more familiar **complete flowers**. Nevertheless, they are fully equipped to play their role in the reproductive stage of the rice plant's life.

Like all green plants, a rice plant reproduces sexually, and its flowers contain male and female reproductive organs. The male organ of a flower is the **stamen**. Its enlarged tip, called an **anther**, produces **pollen**, which contains male reproductive cells, or sperm.

The female part of the flower is the **pistil**, which is made up of three basic parts: a structure called a **stigma** at the top, which is connected by the stalk-like **style** to a chamber known as the **ovary**. Produced within the ovary are tiny **ovules**, which contain egg cells—the female reproductive cells.

32

Above: A rice flower enclosed by the lemma and palea. *Right:* This photograph shows a flower with its surrounding scales removed.

A rice flower has all these basic reproductive parts: six stamens with large anthers and one pistil that includes a single ovary with two styles and two feathery stigmas at their tops. What it does not have are the graceful, showy petals of many other kinds of flowers. Instead of being surrounded by a corolla of petals, a rice flower's stamens and pistil are enclosed in two green, scale-like structures, one overlapping the other. The outer scale, called the **lemma**, and the inner one, the **palea**, lock together, sealing the rice flower inside.

33

Rice plants, like all members of the grass family, are pollinated by the wind.

A rice flower does not need fragrant, colorful petals because, like all grass flowers, it is **pollinated** by the wind. Roses, morning glories, and many other flowers depend on insects to carry pollen from flower to flower. Their vivid petals attract butterflies and bees and provide a landing platform for the insects. Such attractive accessories are not required for pollination in rice and other grasses. All that is necessary is that the reproductive parts of the flowers be exposed to the wind when they are fully developed.

This happens for a very brief period when the rice flower blooms. The lemma and palea separate like two parts of a shell, allowing the anthers on their long, thin stalks to emerge. The pollen grains that they bear are lifted up by the wind, usually settling on the feathery stigmas of the same flower or of other flowers on the

This series of photographs shows the blooming of a rice flower. 10:00 A.M.—the lemma and palea begin to open. 10:30 A.M.—the stamens emerge. 11:00 A.M.—pollen is released from the stamens. 12:00 P.M.—pollination is completed, and the flower closes.

same plant. (Unlike many kinds of flowering plants, most cultivated rice is **self-pollinating**.)

The flowers of a rice plant bloom in order from the top of the head to the bottom, and each flower is open for only about an hour. At the end of this time, the lemma and palea close again, but now a great change has taken place inside the flower. Pollination is completed, and **fertilization**, the next step in the reproductive process, has begun.

An enlarged photograph of pollen grains on a rice-flower stigma.

Fertilization takes place when pollen grains on the flower's stigmas split and send out tiny tubes that extend down the style to the ovary. Sperm cells move down the pollen tubes, and one of them unites with an egg cell in the single ovule within the ovary. When the two cells join, the ovule begins to develop into a seed containing an embryo, endosperm, and all the other parts needed to create a new rice plant.

As the single seed begins to grow, the flower ovary also goes through a transformation. It enlarges, and its walls become dry. The ovary has changed into the **fruit** of the rice plant, the part of the plant that encloses and protects the seed.

Like all grasses, rice has a dry fruit very different from juicy,

A seed develops inside the fruit or grain of a rice plant. These photographs show the grain with the two enclosing scales removed.

many-seeded fruits like tomatoes and apples. The rice fruit, like the fruit of wheat, corn and oats, is known as a **grain**. In all these plants, the several layers of the fruit wall are joined to the seed coat so that the grain and seed make up a single unit. The grain of the rice plant has another covering, the two scales that originally enclosed the rice flower. These scales stay around the grain as it grows, becoming the tough hull of the rice grain.

The fruit of a rice plant is different from a large, juicy fruit like an apple, but it plays the same role in the life of the plant.

As the rice grains ripen, the heavy heads of the plants begin to droop, and the leaves and stems become yellow and dry. Flocks of birds come to the paddies to feed on the ripe grain.

Locusts and other insects feast on the dry leaves of the rice plants.

Left: A worker cutting rice plants with a small mechanical harvester.
Right: This large combine harvests the rice and threshes it.

When the rice grains begin to ripen on the plants, growers drain the fields by opening gates in the levees. By the time the rice is completely developed, the fields will be dry enough so that they can be harvested.

In many rice-growing countries, the rice plants are harvested by workers who use sharp knives to cut down the dry stalks. In other areas, machines are used to cut the plants and to make them into bundles. North American rice growers usually harvest their crops with large combines that cut the stalks and **thresh** the grain—separate it from the rest of the plant—right in the fields.

After the rice is harvested, it must be dried before it can be stored. The grain may be dried by heated air, or it may simply be left in the warm sun until the kernels have lost all their moisture.

Harvested rice plants drying on racks in Japan. Each rice-growing region in this Asian country has its own distinctive kind of drying rack.

These grains of paddy rice were harvested from a single rice plant.

After the harvesting, threshing, and drying are finished, what is left is what growers call **rough** or **paddy rice**. The grains are still covered with the hulls that are the dried remains of the two scales that originally enclosed the rice flowers. These hulls must be removed before the rice is ready to be used as food.

Small rice farmers in Asia and other parts of the world often remove the hulls from their rice by pounding it or grinding it between stones. Commercial rice growers use large shelling machines to do the same job. The finished product is **brown rice**, which lacks the hulls but is still enclosed in the layers formed by the walls of the fruit and the seed coat.

These brown-colored **bran layers** contain vitamins and minerals, and brown rice makes a nourishing food. The bran layers also contain oil, which can cause spoiling if the rice grains are stored for a long time. For this reason and for the sake of appearance, most rice processors mill or polish the brown rice in machines that rub off the outer layers and some of the endosperm underneath. Milling produces the pearly white grains that most people think of as rice.

Left: A shelling machine removes the hulls of paddy rice, producing brown rice. *Above:* Brown rice is transformed into milled white rice by processing in a milling machine.

Milled white rice is made up primarily of carbohydrate, which is filling and easy to digest. To make white rice a more complete food, processors sometimes coat the kernels with substances containing vitamins, which restores some of the food value lost when the bran coats were removed.

Another method of making rice more nutritious is **parboiling** it when it is still in the rough stage. Parboiling involves soaking paddy rice in water and then steaming and drying it before milling. Parboiled rice contains more vitamins and minerals than regular white rice.

This photograph shows three common forms of processed rice: brown rice (*top*); milled white rice (*left*); parboiled rice (*right*).

Rice is unique among the important grain plants because it is the only one that is almost always cooked and eaten as a whole grain. Wheat grains are usually ground into flour, which is then mixed with other ingredients and cooked to make bread. People eat rice grains in their natural forms, except, of course, for the removal of the hull and bran coats. The tiny grain that plays such an important part in the life of the rice plant also has a vital role in nourishing millions of hungry people around the world.

GLOSSARY

anther—the enlarged tip of a stamen where pollen is produced

blade—the flat upper part of a grass leaf

bran layers—layers of material surrounding a rice grain under the hull, formed by the seed coat and the walls of the fruit

brown rice—rice with the hull removed but with the bran layers still in place

coleoptile (ko-lee-AHP-t'l)—a sheath-like structure that protects the developing shoot of a monocot plant

complete flowers—flowers that have four main parts: a calyx, made up of sepals; a corolla of petals; stamens; pistils. Grass flowers are considered incomplete because they lack a calyx and corolla.

cotyledon (kaht-l-EED-un)—a leaf-like structure that nourishes a young plant until its leaves can begin manufacturing food

dicots (DI-kahts)—plants with two cotyledons

dikes—low walls of dirt used to keep water on a flooded rice paddy

embryo (EM-bree-oh)—the part of a seed that will develop into a new plant

endosperm (EN-dih-sperm)—the material in a seed that provides nourishment for the embryo and young plant

fertilization—the union of male and female reproductive cells

fibrous (FI-brus) root—a root system that is made up of many fine, branching roots

fruit—the part of a plant that encloses and protects the seed

grain—a dry, one-seeded fruit in which the seed coat is fused to the fruit walls

heads—the flower-bearing parts of a rice plant

hull—the hard coat surrounding a rice grain, formed by the two scales that enclosed the flower

lemma (LEM-muh)—the outermost of the two scales that enclose a rice flower

levees (LEHV-eez)—low dirt walls used to keep water on a flooded rice paddy

monocots (MON-uh-kots)—plants with one cotyledon

nodes (NOHDS)—the points on a plant stem from which leaves grow

ovary—the chamber at the base of a flower pistil where seeds develop

ovules (AHV-yuls)—tiny structures in the ovary that develop into seeds

paddies—flooded fields where rice is grown

paddy rice—rice grains with the hulls still attached

palea (pay-LEE-uh)—the innermost of the two scales that enclose a rice flower

panicles (PAN-ih-kuhls)—the flower-bearing parts of the rice plant

parboiling—a method of soaking and steaming paddy rice in order to retain some of the vitamins in the bran coats

photosynthesis (fot-uh-SIN-thih-sis)—the process by which green plants use the energy of the sun to make food

pistil—the female reproductive organ of a flower

pollen—a powdery substance produced by a flower's anthers, containing male sperm cells

pollination—the transfer of pollen from the anthers of a flower to the stigma of the same flower or to another flower of the same species

rough rice—rice grains with the hulls still attached

seed coat—the protective outer layer of a seed

seedlings—young plants ready to be transplanted

self-pollinating plant—a plant that reproduces by means of the transfer of pollen from the stamens of one flower to the pistil of the same flower or to another flower on the same plant

sheath—the lower part of a grass leaf that grows wrapped around the stem

shoot—the developing stem and leaves of a plant embryo

spikelet (SPI-klet)—one of the small branches that make up the head of a rice plant

stamen (STAY-muhn)—the male reproductive organ of a flower

stigma (STIG-muh)—the part of the pistil that collects pollen

style—the stalk that connects the stigma and the ovary

tap root—a root system made up of one main root and smaller secondary roots

thresh—to separate grain from the rest of the plant

tillering—the process of growing by the production of tillers, characteristic of many members of the grass family

tillers—secondary stems that spring from the main stem of a grass plant

INDEX